Should Schools Reopen?
Interim Findings and Concerns

Draft Document for Public Discussion

The Independent SAGE Report 2

by The Independent Scientific Advisory Group for Emergencies
(SAGE)

May 22, 2020

Imprint

Should Schools Reopen? Interim Findings and Concerns

Draft Document for Public Discussion

The Independent SAGE Report 2

The Independent Scientific Advisory Group for Emergencies (SAGE)

Submitted to The UK Government and the People of Great Britain & Northern Ireland by Sir David King, former Chief Scientific Adviser, UK Government, Chair of Independent SAGE

May 22, 2020

Published in the UK, 2020

www.independentSAGE.org
@independentSAGE
YouTube: IndependentSAGE

DOI: 10.5281/zenodo.3935529

ISBN print: 978-1-906496-76-0 | ISBN ebook: 978-1-906496-97-5

Technical production: indie_SAGE Technical Publishing Working Group #iSTPWG
https://github.com/consortium/iSTPWG

What is the Independent SAGE?

'We are following the science' is the message the British public have been hearing from government since COVID-19 mitigating measures began. It says it is following the advice of the Scientific Advisory Group for Emergencies (SAGE). But the activities of the committee have been kept secret and excluded from scrutiny by the public or wider scientific community.

In response, on Monday May 4, the Independent SAGE convened as a group of preeminent experts from the UK and around the world. The aim of the Independent SAGE was and is to provide robust, independent advice to HM Government with the purpose of helping the UK navigate COVID-19 whilst minimising fatalities.

The Independent SAGE is chaired by former HM Government Chief Scientific Advisor Sir David King and draws on a range of international and British experts.

What did the Independent SAGE discuss?

The Independent SAGE investigated seven key points to make constructive recommendations to support the government's handling of the coronavirus pandemic in the UK:

1. What are the criteria for lifting lockdown?

2. How can testing and tracing be successfully achieved?

3. Are the policies on quarantine and the shielding of vulnerable groups sufficient?

4. What social distancing policies will be required in the future? In particular, on school closures and public gatherings

5. What port and border controls are required?

6. Where does responsibility lie for managing the pandemic? Are all levels of government being utilised effectively, is the response by central government delivered vertically failing?

7. The UK has enormous globally recognised untapped resources that should be called on, such as the Royal Society and Academies, manufacturers, etc. How do we bring these into the picture much more effectively? Will we manufacture the vaccines we and the world needs in the UK?

In response to these questions, the Independent SAGE assessed measures to date, asked whether the strategy has been effective to date and reviewed what approached they believed should be adopted to minimise deaths, ending the pandemic as soon as possible.

The approach of the committee is to provide a set of recommendations from this first meeting. It will be based on minimising deaths and ending the epidemic as soon as possible.

What will the outcomes be?
The Independent SAGE will provide a series of evidence-based recommendations for the UK government based on global best practice.

What is the format?
The Independent SAGE will engage in an open debate on the topics on the agenda. This evidence session was live streamed on Youtube so the public can see the evidence presented and understand the debate within the scientific community on the most appropriate course of action for the UK government.

Members of the Independent SAGE committee

CHAIR

Sir David Anthony King
Former Government CSA; founder and Chair, Centre for Climate Repair at Cambridge; Senior Strategy Adviser to the President of Rwanda; Chair of Independent SAGE

Members

Professor Gabriel Scally
President of Epidemiology & Public Health section, Royal Society of Medicine

Professor Allyson Pollock
Co-director of the Newcastle University Centre for Excellence in Regulatory Science.

Professor Anthony Costello
Professor of Global Health and Sustainable Development, University College London; former Director at WHO

Professor Karl Friston FRS
Computational modeller and neuroscientist at UCL in charge of developing a generative SEIR COVID19 model

Professor Susan Michie
Professor of Health Psychology and Director of the Centre for Behaviour Change at University College London, member of SPI(B), SAGE sub-committee

Professor Deenan Pillay
Professor of Virology UCL, former SAGE member

Professor Kamlesh Khunti
Professor of Primary Care & Diabetes, University of Leicester

Professor Christina Pagel
Mathematician and professor of operational research, UCL

Dr Zubaida Haque FRSA
Interim Director, Runnymede Trust

Professor Martin McKee
Professor of European Public Health at the London School of Hygiene and Tropical Medicine

Dr Alison Pittard
Dean of the Faculty of Intensive Care Medicine, UK

+++

In addition, there is also:

The Independent SAGE Behavioural Advisory Group

A new Behavioural Advisory Group has been formed to contribute to the discussions, advice and papers of Independent SAGE.

The formation of the group recognises the central role of the behavioural and social sciences in the response to Covid-19. Every measure to counter the spread of infection is dependent upon the understanding, engagement and adherence to guidance of the public, whether that be self-isolating, social distancing, practising self-protective behaviours such as hand cleansing, getting tested or (eventually) getting vaccinated.

In order to gain the adherence of the public to pandemic control measures, it is critical that the authorities are trusted. For this to happen, both central and local authorities must involve all sections of the public as partners in all stages of the development, delivery and assessment of policies.

Equally, it is crucial to consider the impact of every policy and practice on practices on all sections of the public, especially those who are most vulnerable, with a view to reducing inequalities. Diversity is central to the thinking, procedures and membership of our new group.

Our focus, however, is not limited to the behaviour of the public. We will also address that of policy-makers, planners, and health and social care workers. They are all part of a complex system of people and organisations needed to address the problem. Getting the right policies involves understanding the capability, opportunity and motivations of people in all parts of society.

The Independent SAGE Behavioural Advisory Group includes leading authorities in anthropology, behavioural sciences, criminology, health studies and psychology. Some of the members also participate in UK-wide and Scottish government advisory groups. Its role is complementary, not an alternative, to these other roles. It will support Independent SAGE in looking in depth at key policy issues, including ones not currently on the government's agenda.

The Behavioural Advisory Group aims to help in the development of constructive proposals and policies that will help the UK Government lead the country out of the worst public health crisis of our lifetime.

Appendix 1: Members of the Behavioural Advisory Group

Imran Awan is Professor of Criminology at Birmingham City University and a leading expert on Islamophobia and countering extremism. He is a participant in the UK government's SPI-B group.

Val Curtis is Professor of Hygiene and Director of the Environmental Health Group at the London School of Hygiene and Tropical Medicine. She is an evolutionary anthropologist specialising in hygiene behaviour globally. She is a participant in SPI-B.

John Drury is Professor of Psychology at the University of Sussex specialising in research on collective behaviour including behaviour in emergencies. He is a participant in SPI-B.

Susan Michie is Professor of Health Psychology and Director of the Centre for Behaviour Change, University College London. Her research focuses on behaviour change in relation to health and the environment. She is a participant in SPI-B and advises the World Health Organization on behavioural science in relation to Covid-19.

Ann Phoenix is Professor of Psychosocial Studies at the Thomas Coram Research Unit, UCL Institute of Education. Her research focusses on racialised and gendered identities, family lives and home, migration and transnational families.

Stephen Reicher is Professor of Psychology at the University of St. Andrews. His work addresses group processes, with special emphasis on processes of leadership, influence and collective action. He is a participant in SPI-B and in the Advisory Group to the Scottish Chief Medical Officer on Covid-19.

Elizabeth Stokoe is Professor of Social Interaction at Loughborough University. She is an expert in conversation analysis. She is a consultant for SPI-B.

Robert West is Emeritus Professor of Health Psychology at University College London and Editor-in-Chief of the journal *Addiction*. He specialises in addiction and behaviour change. He acts as advisor to Public Health England on tobacco control and behaviour change. He is a participant in SPI-B.

Laura Bear is Professor of Anthropology at the London School of Economics. Her current research focusses on the impact of the Covid-19 pandemic on vulnerable communities. She is a participant in SPI-B.

Table of Contents

1. INTRODUCTION

UK Cabinet Office Minister Michael Gove has re-asserted the Government's position that schools should reopen from 1st June if certain conditions are met. But this has provoked a mixed response with considerable questions being raised by parents, headteachers, teaching unions, local authorities and health professionals. Many Local Authorities have come forward saying they are not ready, and the British Medical Association and teachers' unions are urging caution.

While there is no dispute that schools play a fundamental role in the development of children's emotional, social and intellectual development, it is also important to remember that schools are embedded within communities. The issue of schools reopening during COVID-19 does not just have implications for pupils; it also has knock-on effects for adult staff, parents and the communities and locality from which pupils come from.

That said we recognise the issues facing decision-makers are complex, with the task of balancing numerous, different and sometimes conflicting needs of children, parents, and teaching and school support staff. We understand that there is an imperative for children to return to school for their own wellbeing, and that this will also enable *some* parents to return to work (others will clearly have to remain at home if there is no provision for the children to go to school), but it is also vital that an appropriate level of safety for children, staff and the wider community is ensured.

Using the frameworks of the recently published guidance from UNESCO (new guidelines to provide a road map for safe reopening of schools) and WHO guidance for schools, we have considered (and are continuing to consider) the impact of school opening on children, staff, and the wider community - including parents, grandparents and guardians. Schools do not operate in a social vacuum, and what happens in schools will have wide ramifications for everyone within and outside of schools. It is for this reason that our approach and analysis in this report about whether schools should reopen on 1 June 2020, has to a great extent been led by questions and concerns sent to us by parents, teachers, inspectors, health professionals and ordinary members of the public about the important issue of schools reopening in a few weeks. We were taken aback by the level of knowledge and understanding among ordinary members of the public about the prevalence and transmission rates of COVID-19 in their local communities, and among children and adults, but at the same time it was apparent that the public did not feel that they had sufficient scientific, social, and educational information from the government about the impact of schools reopening on their children, teachers and the wider community.

Questions and concerns from the public varied from whether it was safe to re-open schools, what criteria needed to be in place for schools to reopen, whether schools reopening for some children would sufficiently address concerns about educational gaps and children's welfare, and the implications of schools opening on local communities, extended families and vulnerable children and adults. We were particularly struck by fears about the impact of coronavirus on children and extended members of households among black and ethnic minority parents who were acutely aware of the disproportionate hospital illness and deaths with COVID-19 among particular BME communities.

Schools represent a major axis of local communities. In the same way that school closures were highly disruptive, then school opening, and the manner in which this is done, will also be disruptive. A staggered school opening based on year groups has major implications for childminders, parents and guardians who may have responsibilities for siblings of different ages. The risk of top down decision-making without community involvement will mean that the burden will fall on individuals to find a way to make this work. We strongly recommend that decisions on school opening be made at local level, involving all stakeholders, to ensure there is support available as schools progress to full function.

We make recommendations based on existing evidence and new modelling, recognising the lack of evidence for many issues, and in such cases identifying options which are likely to minimise risk. We are sharing this document for public consultation via an online meeting on Friday May 22 from 12.00 to 1.30pm.

Over the next week we will take on board further public concerns, consult with other experts on education and public health, and review any new evidence from other countries alongside latest data from the UK to update and refine the recommendations in this brief report.

2. TRANSMISSION RISK

2.1 School opening

We believe that decisions on school opening should be guided by **evidence of low levels of COVID-19 infections in the community and** the ability to rapidly respond to new infections through **a local test, track and isolate strategy. There is no clear evidence that these conditions are met. Until they are it is not safe to open schools on June 1.** Some rural areas might be ready to re-open schools earlier than other places. Estimates of levels of infections must be based on up-to-date real time, detailed, local data on suspected and confirmed cases. To ensure that any local outbreaks are quickly spotted and contained, **we strongly recommend that local test, track and isolate programmes are in place and tested before schools re-open.** In cases where schools reopen where these safeguards are not in place, we suggest alternative testing strategies at the end of this document.

2.2 Are children less likely to be infected than adults?

Studies have shown that between 1% and 5% of diagnosed COVID-19 cases are children, but many children may be undiagnosed because up to a third of infected children never develop any symptoms. Ongoing UK data suggest that children are in fact as likely as adults to become infected and carry the virus. They may be less likely than adults to transmit the virus because, for instance, adults are contagious for longer than children. However, the

impact of placing many children in one place could lead schools to become "institutional amplifiers", if asymptomatic children go unnoticed until an adult becomes symptomatic.

2.3 How sick do children get?

If children get symptoms, these are typically similar to other respiratory illnesses: mild fever, cough, sore throat, sneezing, muscle pain and tiredness. There is scientific consensus that children generally have much milder disease than adults, with a very small number of infected children becoming seriously ill. Concern has been raised that some children might develop a COVID-19 related Kawasaki-type immunological disease that may require critical care, but indications so far are that this is extremely rare.

2.4 Can schools trigger new outbreaks?

It is difficult to assess the true risk of infected children transmitting the virus to other children and adults at school. Where there are ongoing new infections within the community, evidence suggests that re-opening schools could increase the spread of the virus, both in the school and the wider community, perhaps by up to 0.3 on R value. Other evidence from Asia however suggests that school closures have little impact on the rate of transmission. There have been recent reports of an upsurge of cases following reopening of schools in France, South Korea and Denmark, leading to re-closing in some instances. However, this does not necessarily infer transmission within schools but could also be because infections are generally going up in places where lockdowns are eased.

2.5 How much difference does delaying school re-opening make to the chance of a child getting sick?

We have used advanced mathematical techniques to estimate how likely children are to get sick depending on when their school reopens. The **table** below shows the example impact that sending a child to school has on their chance of getting sick with COVID-19, if they went back to a classroom of 15 pupils on June 1st, June 15th, and September 1st. We look at the chances of a child being exposed to an infectious person, the chances of getting sick and the chance of dying from the virus.

These estimates come from mathematical models of the spread of COVID-19 in the UK, based on the most recent national-level data available, and are for young primary school aged children (less than 10 years old). We look at what might happen if a child goes to school (in bold), or stays at home (in italics).

All risks to children are very low and all risks get lower over time as COVID-19 cases become less common (assuming the virus "reproductive number" R remains below 1).

Delaying a school re-opening by two weeks (to 15th June) approximately halves the risk to children, and delaying the re-opening till September is less risky still.

Staying at home at all time points is about half as risky as going to school, but also means that children do not get the benefit of having face-to-face learning and seeing their friends.

To put the very low chances of death from COVID-19 in perspective, the daily chance of being killed in a road traffic accident is about 0.074 per million **(0.07/M)**. So, schools reopening in September present a slightly lower risk and reopening in June a slightly higher risk to a child than the background risk of a road traffic accident. Details of the uncertainty levels are given in the Appendix.

DATE OF RETURN	JUN 1		JUN 15		SEP 1	
	School	*home*	*school*	*home*	*school*	*home*
WHAT IS THE CHANCE A CHILD WILL BE EXPOSED TO A CONTAGIOUS CLASSMATE TODAY?	4.21%	1.76%	2.09%	0.87%	0.49%	0.19%
WHAT IS THE CHANCE A CHILD WILL CATCH THE VIRUS TODAY?	1.46%	0.61%	0.72%	0.30%	0.15%	0.06%
WHAT IS A CHANCE A CHILD WILL DIE FROM THAT INFECTION? (per million)	0.23/M	0.10/M	0.11/M	0.05/M	0.02/M	0.01/M

The risks can be visualised in another way:

1. Going to school

Each large square represents 400 primary school age children (one small square per child).

1st June   15th June   1st September

Out of every 400 non-infected children **starting back at school from** …

Average number of children (orange) exposed to an infectious person

Average number of children (red) who then catch COVID-19

It would be extremely unlikely (less than 1 in a million chance) that a child would die

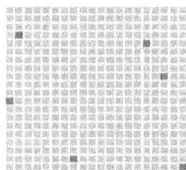

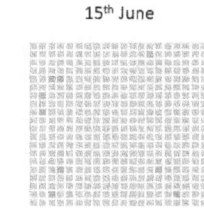

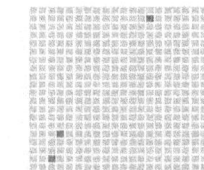

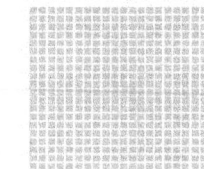

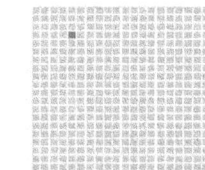

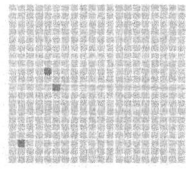

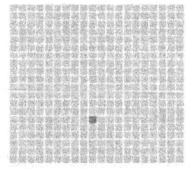

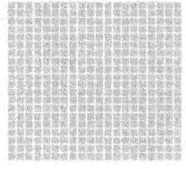

2. Staying at home

Each large square represents 400 primary school age children (one small square per child).

1st June   15th June   1st September

Out of every 400 non-infected children **staying at home** …

Average number of children (orange) exposed to an infectious person

Average number of children (red) who then catch COVID-19

It would be extremely unlikely (less than 1 in a million chance) that a child would die

The main takeaway from this kind of mathematics is that **risk falls relatively quickly over a week or two** after commitment to a June 1 reopening date. This means that **delaying re-opening by a couple of weeks could allow time to find solutions to local challenges and set up strong local testing procedures while knowing that risks are getting lower**.

2.6 What happens if a school has new cases of COVID-19?

Robust testing and tracing procedures along with support for people and families to self-isolate will reduce the chance of infectious staff, parents or children attending school (or anywhere else). They will also quickly spot any new cases of infection that do arise in a school. So if a class or school then has to close temporarily after reopening, this should not then be seen as a failure — or evidence that opening was premature — but instead as an integral part of a community-based tracking and testing programme that will play an essential role in delaying, and hopefully preventing, any second wave of infection. In other words, the school community may have a central role to play in not only meeting the educational and other needs of children but also **in providing an effective surveillance structure** that may be essential in keeping local communities safe.

We need a capability for a **real time response to TTI data**, with clear criteria to act and plans in place to re-close schools if need be. This should not be seen as a failure or a cause for blame. It's inevitable in a pandemic that new outbreaks will occur. Planning for such re-closure is essential and must include measures to maintain educational opportunities for pupils.

2.7. What is the risk to school staff, parents and household contacts?

Even if there are very few new infections within schools, this could still be risky for some adults who come into contact with infected children. This might include teachers and other school staff, household members, childminders, and any other adults the child may have contact with. Most younger teachers who are healthy are unlikely to get more than a mild disease. But we know that factors such as age, being a man, coming from a low income background, underlying health conditions (e.g. diabetes, high blood pressure) and being from black and minority ethnic (BAME) backgrounds may make teachers and staff more vulnerable, in particular in cities with high BAME populations.The risks for those most vulnerable and those shielding are very much higher than an adult without any risk factors.

Many schools with BAME children are based in more deprived areas where the risks will be greater and cumulative. We are well aware of robust evidence (including recent findings from the ONS) which shows that people from deprived areas, as well as particular BME groups (including Bangladeshi, Pakistani and black groups) are significantly more likely to die from COVID-19 than their more advantaged or white British counterparts. Additionally, many key workers are also from lower income and BAME backgrounds enhancing their risk of vulnerability and risk of infection. Since many children (e.g. approx. 30% primary school and 25% of secondary school children are from BME backgrounds), teachers and other school staff are from black and ethnic minority communities, **it is important to consider the locality-based COVID 19 infection and death rates as the best indicator of the risk from any future school-based outbreaks.** We plan to have modelled these effects before we release our full report next week.

3. EDUCATION RISK

3.1 Schools in deprived areas

The Children's Commissioner calls for attention to the wider social and economic costs of keeping schools closed and encourages intelligent, incremental reopening of schools, particularly nurseries and primary schools, responsive to local contexts and informed by rigorous testing and comprehensive data.

Whilst there is little real-world data on the educational impact of school closures, we know that teachers in the most deprived schools are **more than twice as likely** to say that the work their students are sending in is of a much lower quality than normal (15% vs 6%). In the most deprived schools, 15% of teachers report that more than a third of their students learning from home would not have adequate access to an electronic device for learning, compared to only 2% in the most affluent state schools. **12% of those in the most deprived schools also felt that more than a third of their students would not have adequate internet access.**

A recent report from the Institute of Fiscal Studies also highlighted the disproportionate impact on deprived areas. Higher-income parents are much more likely than the less well-off to report that their child's school provides online classes and access to online videoconferencing with teachers. 82% of secondary independent school pupils are offered active help, with 79% being provided with online classes. 64% of secondary pupils in state schools from the richest households are being offered active help from schools, such as online teaching, compared with just 47% from the poorest fifth of families.

Moreover 60% of independent schools and 37% of schools in the highest income areas had an online platform to receive work, compared to 23% in the most deprived schools. 45% of students had communicated with their teachers in the last week. At independent schools, the figure is 62% for primaries and 81% for secondaries.

The decision of many independent schools to open in September demonstrates their ability to prioritise infection prevention, but only in the context of ongoing high quality on-line educational opportunities for their pupils.

3.2 Access to facilities

It is clear that existing stark inequalities in educational opportunities for children are being further widened through the COVID-19 pandemic, in parallel with the clear excess disease burden in the most disadvantaged adults - many of whom are parents and grandparents.The impact goes further than academic progress, including the psychological effects of isolation and increased vulnerability to some children when they are never seen by adults outside their own household.

Returning to school is important for children psychologically and socially, as well as educationally and we should aim to support all year groups to return to school safely. In vulnerable households or where there is a history of domestic abuse it is even more important.

We therefore raise the possibility of using a wide range of empty facilities - private schools, sports facilities as examples - to allow those most in need of face-to-face education and social support to receive it, not only during the summer school term but also over summer holidays. This could be in the form of summer camps where social and physical development would be supported, with some education where possible.

4. INFECTION CONTROL IN SCHOOLS AND COMMUNITIES

4.1 Schools

A wide range of infection control measures have been implemented as schools have reopened around the world. These range from staggering opening by year groups, smaller class sizes, outdoor teaching and variable start times through to personal protection, physical distancing, and regular handwashing including, of course, for teachers and other staff at school.

However, the crucial factor allowing school reopening around the world has been the presence of well-functioning local test, trace and isolate protocols—something that is now accepted will not be in place in England by early June. Each school will vary in terms of layout, ventilation, and numbers of available staff which will determine precisely which infection control procedures are feasible or not. We note that school opening in other countries, such as Denmark, has been preceded by substantial investment in measures such as additional washing facilities to promote safety.

Key measures need to be in place to ensure safety, and each school needs to consult widely with teachers, unions, parents, local authorities including public health for local intelligence on infectivity, education departments and inspectors and children before proposing how best to open the school to the proposed additional full classes.

4.2 Community

We recognise that schools will start to open while social distancing guidance remains in place. Issues which require clear guidance include transportation to school, collection of children at the school gates, often by childminders, grandparents, and other carers, and how staff and children can best minimise potential transmission of infection in households and communities. Consideration will need to be given to family members being from vulnerable groups and being shielded and for multigenerational family households. Schools need to work with the local communities to develop specific protocols suited to the local environment (for instance, a village school faces different challenges than an inner city one). Dealing with children's anxiety about social distancing also requires careful consideration.

5. HOW TO PLAN FOR OPENING SCHOOLS SAFELY AND WITH OPTIMAL EDUCATIONAL SUPPORT?

The following should be in place before a school can open safely:

5.1 A risk assessment

A risk assessment should be conducted at four levels including risk assessment of the school, the staff, the pupil and the parents and family environment. A risk assessment should follow **full engagement with local authorities, school managers, trade unions, parents, inspectors, staff and the community.** The assessment process needs to be understood and tailored to the risks and contexts of communities and their members and to the feasibility of schools to implement appropriate infection control procedures. This includes the different risks from lessons in different subjects (e.g. where there is sharing of equipment) and the ability to socially distance within schools. Consideration should be given to use of non-school facilities which may provide a better infection control environment, such as independent school buildings and playing fields, and sports facilities and football stadiums, which will be unused during this time. Such engagement will ensure an optimal solution to providing high quality education, social interaction and physical activities to children. The ability to be outside or in very well ventilated buildings or marquees will be associated with reduced infection rates.

5.2 Are local infections low enough?

Local communities need to be sure that there are **few people currently infected and that numbers of new infections are decreasing**, with the definition of 'few' considered in the context of declines locally over the previous 2 weeks and numbers at the peak of the pandemic. Such data may not be available; data access and public availability is a critical gap in the current management of the pandemic.

5.3 Hygiene and personal protection

Schools need to ensure adequate access for hand hygiene including sufficient and clean toilets and wash facilities and hand sanitisers. As well as social distancing, some schools in countries such as France and China are making the use of face coverings compulsory.

Children should be given designated equipment including stationary to reduce the risk of contagion spread.

5.4 Find cases, test, trace, isolate.

A **test, trace, isolate (TTI) infrastructure** can rapidly suppress transmission, as well as provide confidence to staff, pupils, families and the local community that there will not be new uncontrolled local outbreaks. In the absence of a top-down, well-functioning system, which is not expected to be in place in the very near future, we recommend **local solutions to TTI, linked to local public health authorities and primary care, but with systems in place to allow as near as possible to for local data to contribute to national data collection**. This could involve wider testing of staff and pupils in schools to quickly identify asymptomatic infections, within an ethical, appropriate and agreed structure.

6. PRESERVING EDUCATION IN THE SUMMER AND IF SCHOOLS STAY CLOSED

6.1 Wifi and online education

A major concern is to provide opportunities for pupils in state schools who may not have access to wifi, computers, tablets or smartphones if education is forced to move largely online. In a digital world we believe every child in school aged seven and above should have access to these facilities. Options for requisitioning independent schools, academy schools, church halls, sports clubs, football stadiums and other facilities should be considered to provide socially distanced facilities for all children.

We must also support children from families who find home schooling difficult for other reasons (such as not having English as a first language).

6.2 Summer camps and open-air education

Local authorities and civil society groups should be mobilised to provide **summer schools and camps** to help with educational "catch up", particularly for those most disadvantaged by the lockdown, and also to provide some respite for parents and carers. This presents a major opportunity for community engagement and potentially the use of some of the 750,000 volunteers.

Use of sports grounds, football, rugby and other stadia and hire of marquees will provide opportunities for exercise and socially distanced education for children. If teachers are not available, trained coaches and approved supervisors would be needed for these activities including music and drama.

Governments have a duty to provide the **investment and resources** for schools and staff, for reduced class sizes, and, where schools do not open or open partially, to take steps to provide alternatives to ensure meaningful education that meets the needs of all children .

6.3 Hidden hunger

We remain concerned about the level of hidden hunger among children from poorer households, with parents who are on benefits or in low-paid employment . According to **UNICEF**, in 2017, 10% of children in the UK were living in households affected by severe food insecurity. According to **figures** published by **End Hunger UK**, in January 2018 16% of adults in Great Britain either skipped or saw someone else in their household skip meals; 14% of adults worried about not having enough food to eat; and 8% of adults had gone a whole day without eating because of a lack of money during the last 12 months.

Provision of midday meals for vulnerable children out of school and during the summer months is essential, supported by government, local authorities or via civil society organisations.

6.4 WHO Check Lists

We suggest that schools consider using the checklists for parents and children and for students as suggested by the Word Health Organisation regarding Actions for COVID-19 Prevention and Control in Schools. See Appendix 3.

APPENDIX

1. National Education Union Criteria for Schools Re-opening

Test 1 : **Much lower numbers of Covid-19 cases** The new case count must be much lower than it is now, with a sustained downward trend and confidence that new cases are known and counted promptly. And the Government must have extensive arrangements for testing and contact tracing to keep it that way.

Test 2 : **A national plan for social distancing** The Government must have a national plan including parameters for both appropriate physical distancing and levels of social mixing in schools, as well as for appropriate PPE, which will be locally negotiated at school-by-school and local authority level.

Test 3 : **Testing, testing, testing!** Comprehensive access to regular testing for children and staff to ensure our schools and colleges don't become hot spots for Covid-19.

Test 4 : **Whole school strategy** Protocols to be put in place to test a whole school or college when a case occurs and for isolation to be strictly followed.

Test 5 : **Protection for the vulnerable.** Vulnerable staff, and staff who live with vulnerable people, must work from home, fulfilling their professional duties to the extent that is possible. Plans must specifically address the protection of vulnerable parents, grandparents and carers.

2. Uncertainty levels for the risk of contagion

The following graphic illustrates the uncertainty that attends the predictions of risk (here, the risk of contagion). This is probably too much detail for the consultation document but could be used to reassure people that a formal risk analysis could be pursued if and when necessary. The graph below shows when we would expect the risk of contagion to fall below 1% (broken line) based upon UK averages. This analysis could be repeated for regional data (e.g., time series of new cases and deaths).

Contagion risk (%)

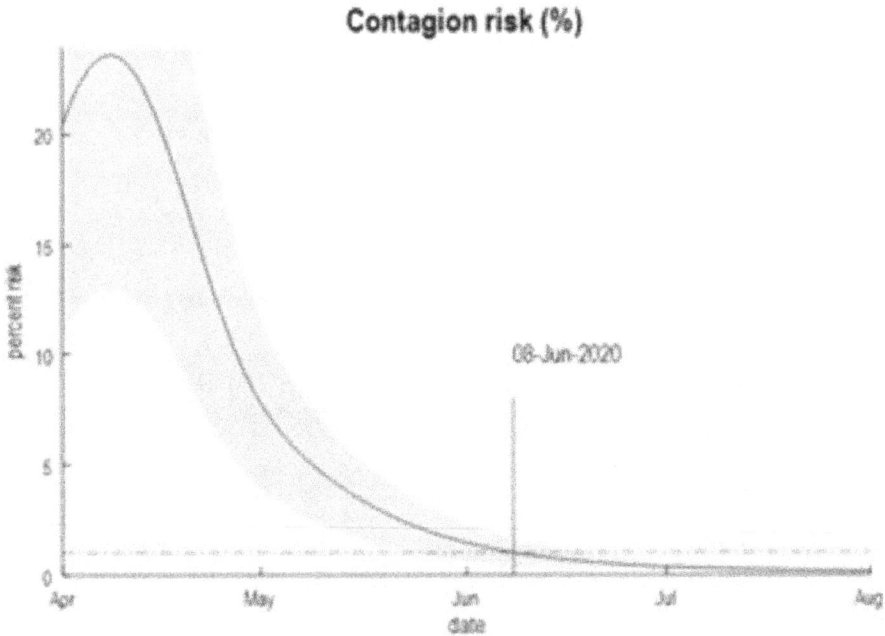

From the point of view of the epidemiology, June 1 stands in for 52 days following the peak of infections (10 April for London). This means 'June 1' is the earliest date at which risk may be acceptable for some areas. **Other areas could defer their 'June 1st' in relation to their local experience**.

3. WHO Checklists

CHECKLIST FOR PARENTS/CAREGIVERS & COMMUNITY MEMBERS

☐ 1. Monitor your child's health and keep them home from school if they are ill

☐ 2. Teach and model good hygiene practices for your children

- Wash your hands with soap and safe water frequently. If soap and water are not readily available, use an alcohol-based hand sanitizer with at least 60% alcohol. Always wash hands with soap and water, if hands are visibly dirty
- Ensure that safe drinking water is available and toilets or latrines are clean and available at home
- Ensure waste is safely collected, stored and disposed of
- Cough and sneeze into a tissue or your elbow and avoid touching your face, eyes, mouth, nose

☐ 3. Encourage your children to ask questions and express their feelings with you and their teachers. Remember that your child may have different reactions to stress; be patient and understanding.

☐ 4. Prevent stigma by using facts and reminding students to be considerate of one another

☐ 5. Coordinate with the school to receive information and ask how you can support school safety efforts (though parent-teacher committees, etc.)

CHECKLIST FOR STUDENTS AND CHILDREN

☐ **1.** In a situation like this it is normal to feel sad, worried, confused, scared or angry. Know that you are not alone and talk to someone you trust, like your parent or teacher so that you can help keep yourself and your school safe and healthy.

- o   Ask questions, educate yourself and get information from reliable sources

☐ **2.** Protect yourself and others

- o   Wash your hands frequently, always with soap and water for at least 20 seconds
- o   Remember to not touch your face
- o   Do not share cups, eating utensils, food or drinks with others

☐ **3.** Be a leader in keeping yourself, your school, family and community healthy.

- o   Share what you learn about preventing disease with your family and friends, especially with younger children
- o   Model good practices such as sneezing or coughing into your elbow and washing your hands, especially for younger family members

☐ **4.**   Don't stigmatize your peers or tease anyone about being sick; remember that the virus doesn't follow geographical boundaries, ethnicities, age or ability or gender.

☐ **5.**   Tell your parents, another family member, or a caregiver if you feel sick, and ask to stay home.